NATURE

ANIMAL HOMES

JOYCE POPE

Illustrated by
JAMES FIELD

Troll Associates

Nature Club Notes

Though you may not know it, you are a member of a special club called the Nature Club. To be a member you just have to be interested in living things and want to know more about them.

Members of the Nature Club respect all living things. They look at and observe plants and animals, but do not collect or kill them. If you take a magnifying glass or a bug box with you when you go out, you will be able to see the details of even tiny plants, animals, or fossils. Also, you should always take a notebook and pencil so that you can make a drawing of anything you don't know. Don't say "But I can't draw" – even a simple sketch can help you identify your discovery later on. There are many books that can help you name the specimens you have found and tell you something about them.

Your bag should also contain a waterproof jacket and something to eat. It is silly to get cold, wet, or hungry when you go out. Always tell your parents or a responsible adult where you are going and what time you are coming back.

If you watch quietly, you can often discover where an animal has its home or den. Do not try to interfere with it. You are much, much larger than most of the creatures that you are likely to see, and can easily frighten them, even though you do not mean to. It is all too easy for you to damage a nest or lead enemies to it. By watching from a distance, you give the animal you are interested in a better chance of survival.

Library of Congress Cataloging-in-Publication Data

Pope, Joyce.
   Animal homes / by Joyce Pope ; illustrated by James Field.
      p.   cm. — (Nature club)
   Includes index.
   Summary: Describes a variety of animal homes, including winter dens, tree homes, and mammal nurseries.
     ISBN 0-8167-2775-9 (lib. bdg.)   ISBN 0-8167-2776-7 (pbk.)
   1. Animals—Habitations—Juvenile literature. [1. Animals—
—Habitations.] I. Field, James, ill. II. Title. III. Series.
QL756.P66 1994
591.56'4—dc20   91-45380

Published by Troll Associates

Copyright © 1994 by Eagle Books

All rights reserved. No part of this book may be reproduced or utilized in any form or by any means, electronic or mechanical, including photocopying, recording or by any storage and retrieval system, without permission in writing from the Publisher.

Designed by Cooper Wilson, London
Edited by Kate Woodhouse

Printed in the U.S.A.

10 9 8 7 6 5 4 3 2 1

Contents

Nature Club Notes 2
Territories and Homes 4
Different Kinds of Territories 6
Water Territories 8
✓ Dens 10
Winter Dens 12
✓ Tree Homes 14
Nests and Nurseries 16
Mammal Nurseries 18
Town Houses 20
✓ Insect Builders 22
✓ Ants and Termites 24
Changing the Environment 26
✓ Animals In Our Homes 28
Glossary 30
Index 31

Territories and Homes

Most people live with their family in a building, which is their home. At home, people feel safe. They can live as they please, and nobody will interfere. Many animals, such as butterflies, do not have a special safe place. They move about their *environment* and find shelter when they need it. Other creatures live in what is called their *home range*. This is usually a large area, in which they find food and other things they need. They share the home range with different animals, but they are often careful not to meet other members of their own species. In the home range there is often an area called the *territory*. The animal defends this against intruders of its own kind, though other sorts of creatures may use it.

▼ The pine marten lives in a hollow tree or a rock formation. It hunts rabbits, squirrels, and birds. The marten will mark its territory by leaving its scent on rocks and branches. This warns other martens that the area is occupied.

◀ This butterfly is seeking shelter from the rain. It does not have a special place in which to take shelter so it lands under any leaf large enough to keep it dry.

An animal recognizes its territory by landmarks, such as trees or stones, and it often marks the boundary with scent messages. Sometimes big animals, such as bears, tear pieces from the bark of trees. This is a clear signal for others that means, "Keep off; this is my living space."

In the spring, birds announce that they have found a territory by singing loudly near its boundary. They also defend the area by display, showing their bright colors. If this does not frighten strangers away, the territory owner may use force.

▲ Birds sing to tell others that they own a particular territory. Some, like the yellow-shafted flicker, use their beaks to drum on hollow branches. The noise attracts mates and keeps rivals away.

Different Kinds of Territories

Most animals look for their first territory when they leave their parents. Often they have to go a long way before they find one. Scientists have discovered that animals that do not get a territory often die. It seems that without the security of a territory, they become ill or get caught by enemies more easily.

There are almost always fewer good territories than there are animals wanting to find one. When a territory owner dies, its place is soon taken by another animal.

Some territories contain all the food, water, and shelter that an animal needs. When some creatures, such as certain kinds of owls, have found a suitable place, they defend it and live there for the rest of their lives.

▼ Hillstar hummingbirds live high in the Andes mountains. They defend their food plants, so there is always enough nectar for them.

If food is difficult to find, some animals have two territories, one for living and one for feeding. A pair of hummingbirds, for instance, often defends a small patch of flowers that provides food for them and their chicks. It may be some distance from the home area which contains their nest.

Sea birds often nest a long way from the places where they feed. Most kinds *rear* their young in dense colonies. Each nesting territory is so small that a bird sitting on its eggs can just about reach its nearest neighbor.

▼ These gannets are courting on their nest site. Their territory is a densely packed breeding colony.

Water Territories

Animals of the open sea do not defend territories, but those that live near the shore or in fresh water often do. These territories, as on land, may be used for a short time while a family is reared, or they may be a lifelong home. Just as on land, each species living in an environment has its own special needs, so the territories overlap. As a result, the whole *habitat* is used as fully as possible.

▼ A coral reef is home to many creatures.

One of the best-known territorial water animals is the three-spined stickleback, a small fish. In the spring or early summer, the males choose sheltered territories for making a nest. The males change color, their undersides becoming bright red. Each male defends his territory by displaying the color to his neighbors, and by showing the sharp spines on his back. When he is at home, the male is the boss in his own territory and he will attack anything red. However, if he moves into another territory, he allows himself to be set upon by its owner and tries to get back to his own living space. This behavior, so different in and out of the territory, is like that of all territorial animals.

Females lay their eggs in the nest that the male has made. The male cares for the eggs until they hatch, and for a short time he looks after the baby fish. After this, his color fades and he joins a group of other sticklebacks until the next spring.

▼ A male three-spined stickleback will defend the territory in which he has made a nest by showing his bright colors and his sharp spines to rival males.

crown of thorns starfish

damsel fish

large sea anemone

Dens

Most animals have a special safe place within their territories where they can sleep and rear their families. This is called a *den* and is often underground. Usually a den has more than one entrance. If any enemy comes in the front way, the den owner can escape through the back door.

Sometimes a den is just a simple shelter. Some animals, such as weasels, make several dens in different parts of their territory and move between them, never staying long in one place. More often, a den is a permanent home. It has a sleeping area and at least one place where food is stored. Shrews and moles, which eat grubs and worms, take any surplus back to their dens. A scientist once found a mole's den with over 400 worms stored in it.

▼ A mole's den is called a fortress. A mole will hunt in the long tunnels that lead away from the fortress. It returns to the fortress to store surplus food and to rest.

▲ European badgers make elaborate dens, called setts, in hills. Several animals share the sett, which may have tunnels that are over 328 feet (100 m) long. There are warmly lined sleeping chambers.

The sleeping chamber is often lined with leaves and dried grass, which makes warm bedding. Sometimes this is woven into a neat nest. Badgers collect a great deal of bedding and throw it out when it becomes soiled. In the springtime, they may take sap-covered leaves to add to their bedding. As these leaves begin to rot, they heat up and make a warm covering for the cubs.

Winter Dens

Some animals do not use permanent dens during warm weather. In the summer, brown bears, for instance, may travel long distances to find food and do not make elaborate dens at every stopping place. They sleep in the shelter of dense trees. As winter approaches, they look for a place protected by rocks or among the roots of a large tree. They dig a secure den, line it with leaves, and snooze until springtime. Females often make a deeper shelter than males, since they give birth to their cubs in the depths of winter and must have a safe place for them for a few months.

▼ The female polar bear digs a den in a bank of snow, where she will give birth to her young. The den protects the newborn cubs from the freezing winds and storms. The cubs do not emerge from the den until they are about three months old, at the start of the Arctic spring.

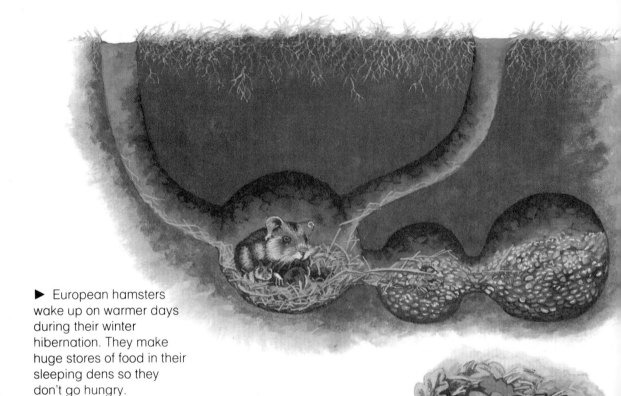

▶ European hamsters wake up on warmer days during their winter hibernation. They make huge stores of food in their sleeping dens so they don't go hungry.

Bears can wake up and escape quickly if they are disturbed. Many other creatures go into complete *hibernation*, which is more like being in a coma. They need a safe place to spend the winter, for they cannot rouse themselves easily. Marmots and similar animals burrow underground to soil that remains unfrozen. They usually disguise their hibernating place by making side burrows. They block the main tunnel with the earth from the side burrows. A meateater that found the entrance would think it was a dead end.

Hibernating animals carry bedding to their dens and make huge stores of food in case they break their winter sleep. They also need food for the lean spring season, when there isn't much for them to eat.

▲ Some hedgehogs hibernate during winter by rolling themselves up in a ball of dead leaves and grass. They do not store food, but live off the fat that they have developed during the summer.

Tree Homes

Some kinds of animals live in trees, where they find food, shelter, and a place to bring up a family. A big tree is like a city, teeming with life. Insect grubs live in the wood, under the bark, and among the leaves. Many birds and mammals that rarely go down to the ground eat the insects. Others eat nuts and fruit produced by the trees.

Old trees often have broken and hollow branches. These make good shelters for many creatures such as raccoons, that need protection from enemies and bad weather. Some birds use hollow trees for their nests, and woodpeckers use their strong beaks to chisel out new holes each year.

▶ Hollow trees make good shelters for some kinds of bats.

▼ During winter, gray squirrels will leave the nest to look for nuts that they buried in the autumn.

Twigs and branches are often used to make shelters. In the summer, squirrels make their nests from leafy twigs on the outer branches. In the winter, squirrels retreat to larger nests built against the tree trunk, where they will be protected against the wind. A winter nest may be used by generations of squirrels, each adding soft grass and moss to keep it cozy and windproof.

Gorillas, chimpanzees and orangutans make comfortable resting places for themselves in the trees. Wherever they stop to sleep, they break and weave branches to make mattress-like platforms strong enough to hold their weight. They leave these platforms behind when the group moves on the next day.

▼ Chimpanzees use broken branches and leaves to make springy beds high above ground level.

Nests and Nurseries

Many animals make special nests to rear their families. Birds' nests are best known, but some fish, reptiles, insects, and mammals also make *nursery* nests. These are often elaborate, but are used only for a short time. Some long-lived birds, such as eagles, may return to the same nest for many years. They leave it once the young eagles can fly and return when they are ready to rear another family of chicks.

Birds use many different materials to build their nests. Most sea birds and waders just make a slight scrape in the gravel or sand of the shore. Adélie penguins lay their eggs on piles of pebbles, while many ovenbirds make mud structures so solid that they last for years. Swiftlets make nests of saliva that hardens. The smallest nests are made by hummingbirds. The largest are communal nests made by some weavers, which seem to fill the small trees in which they are built.

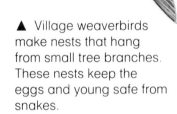

▲ Village weaverbirds make nests that hang from small tree branches. These nests keep the eggs and young safe from snakes.

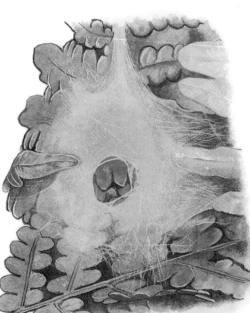

◀ This tangle of silk threads is the entrance to a spider's den. A silk-lined tube leads from the entrance to a safe, warm living space.

▶ The South American ovenbird builds a nest of mud strengthened with twigs.

▲ Some swifts use their own saliva to glue bits of straw or feathers together to make a nest. Others like the swiftlet above only use saliva to make a hard cup in which eggs are laid.

▲ Many songbirds, such as the bullfinch, make a deep cup-shaped nest in which their eggs and young are kept warm and safe.

▲ Mound building birds do not sit on their eggs, but use various kinds of natural heat to keep their eggs warm.

Most songbirds produce helpless young, which are protected in nests woven of twigs or grass and warmly lined with moss or feathers. The nests range from the deep cup made by finches to the beautiful hammock of the Baltimore oriole.

Megapodes (which means "big feet") such as the mallee fowl of Australasia, do not make a nest, but gather a mound of dead plants and sand. The female bird lays her eggs in the mound, and the male guards it and keeps the rotting vegetation at the right temperature.

Mammal Nurseries

Large mammals, such as elephants and antelopes, do not need to make safe nests for their families. Their babies are born well developed and are active within minutes of their first breath. One large mammal to make a nest is the wild boar. The sow digs a shallow trench and lines it with dead leaves. Perhaps because of this she gives birth to more young than any other hoofed animal.

Some nurseries are no more than simple shelters. Female bats gather in nursery roosts to give birth in early summer, while the males live elsewhere. Animals that develop slowly, such as pine martens, are usually born in nests on the ground.

▶ The harvest mouse builds a nest among stalks of grass. When the grasses die in the autumn, harvest mice move to the protection of hedge banks or scrub land.

▼ Wild boars make a special nursery for their young. The piglets do not remain in the nest for long and soon follow their mother when she leaves to hunt for food in the forest.

Small mammals often make a special nursery for their young. Old World rabbits dig a short burrow. *Dominant* females make theirs in safe places near the middle of the *warren*. Others have to care for their young near the edge of the colony, where there is more danger from *predators*. The nests are lined with soft grass and fur that the mothers pull from the underside of their bodies.

One of the most beautiful of all mammal nests is made by the harvest mouse. It shreds growing leaves on the stems of large grasses and uses them as the framework for the nursery, which is lined with soft plants. The nest is difficult to see because the harvest mouse hides its nest among tall stalks of grass.

▼ Adult martens are wonderful climbers, but their young cannot climb when they are first born. Because of this, they are usually born in a den hidden among rocks on the ground or a cliff.

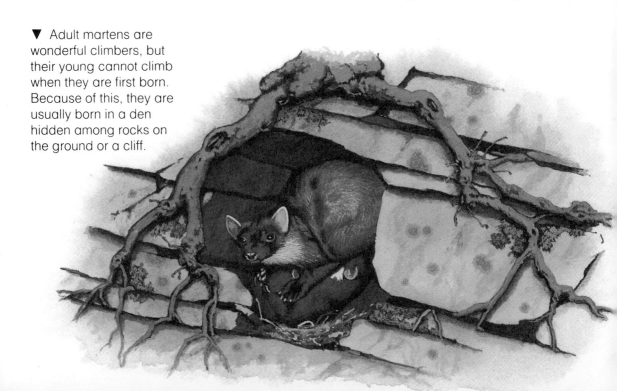

Town Houses

Some mammals, particularly those that eat plants, live in large family groups. Sometimes many families gather together. Huge herds of antelopes and zebras still travel across protected parts of Africa. But even larger numbers of a few burrowing species can be found herded together below ground.

The prairie dogs of North America were at one time the most numerous of all animals living together in one home, but they are now far less common. They are not dogs, but ground squirrels. Their colonies are called towns. The largest of these, in western Texas, had a population of 400 million prairie dogs and covered nearly 25,000 square miles (65,000 square km). This huge area included valleys and hills, which divided the town into wards. Each ward had a number of smaller groups called coteries. The coteries acted as family units and defended their living space. The whole colony lived peacefully. Many other creatures lived in the burrows, so the town was a complex community.

▼ The burrows of prairie dogs provide homes for many other creatures. Some such as the black-footed ferret and the rattlesnake live in the shelter of the town and prey on its members.

Burrowing animals are safe from many enemies. To prevent their numbers from becoming too large for the food supply, some burrowers, such as the African naked mole rat, reduce their families by allowing only one female to bear young. The rest of the mole rats work to enlarge the burrows, gather nest material, and search for food for the whole family. When the chief female dies, one of the others takes her place as "queen."

▲ The young of the naked mole rat queen become workers at first, enlarging the burrow and collecting food. Later, some of them become non-workers. They live with the queen and are supported by the rest of the family.

▼ Prairie dogs have many advantages by living in groups. One is that predators are more likely to be spotted so that the town can be alerted and avoid being preyed upon.

Insect Builders

Very few insects have what we would call a home. Most of them live anywhere they can find food. Some caterpillars protect themselves by spinning silk webs around the place where they are feeding. More permanent homes are made by the few insects that care for their young. Female earwigs, for example, look after their eggs and newly hatched young in an underground cell. But the biggest insect families are the social insects – ants, termites, and some bees and wasps.

▼ Over 70,000 bees can live in a single hive. Most are workers who cannot lay eggs but do all the cleaning, building and feeding the young grubs. In the summertime, a busy bee lives for about six weeks, though a few bees survive in the hive throughout the winter.

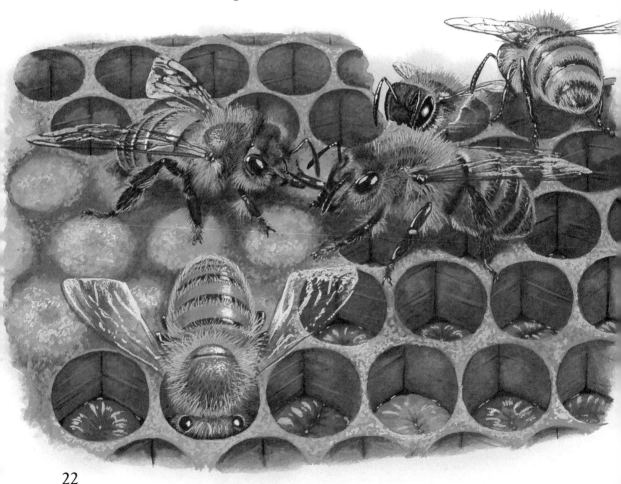

▲ A worker wasp feeds the grubs in her nest. Wasps lead an upside-down life, because the paper cells in which the grubs develop open downward.

The homes of honeybees and social wasps are made of six-sided cells. Bees' cells are made of wax, while worker wasps scrape and chew dead wood to make paper for their nests. Wasp nests and bumblebee hives are nurseries, in which *grubs* are reared by the workers. Honeybee hives have storage cells to hoard honey for the winter.

Most of the inhabitants of these nests are workers who build and keep the nest clean, find food, and bring it to the grubs. The workers also protect their home. They are able to do this with their egg-laying tube, which cannot produce eggs but has developed into a stinger.

Social wasps and bumblebees die after a new generation of queens has been reared. Honeybee colonies survive, because although the life of the workers is short, the queen may live for several years and produce thousands of young each summer.

▲ The caterpillar hunter wasp makes a neat clay pot and stocks it with paralyzed caterpillars. When the pot is full, she lays an egg. The grub that hatches from the egg feeds on the caterpillars.

Ants and Termites

Ants and termites look somewhat alike, and termites are sometimes called "white ants." But these insects are not closely related.

All ants live in large groups, and most make nests to shelter the queen, workers, and young. An ant nest looks untidy and haphazard compared with the precision of a wasp nest or beehive, but it has advantages. If an ant nest is attacked, the workers can easily move eggs, grubs, or *pupae* to safety. It is like a house in which the rooms can be adapted to all sorts of uses.

Some ant nests are very large, and are home to hundreds of thousands of workers. The ant workers may have a long life, and a queen ant can survive for twenty years. Some ants cultivate fungus gardens underground for food.

▶ Ants carry their pupae to safety when their nests are attacked.

▶ Leaf cutter ants carry bits of leaves to their nest. They are used to make a compost heap on which special fungus grows, providing food for the ants.

▲ Some kinds of queen termites grow so large that they cannot move. They are cared for and fed by workers.

Termite nests can also be large. Often a big part of the nest is underground. The largest nests have huge towers, which mainly control air flow and temperature. At one time, when human architects were planning very tall buildings, they studied termite nests to see how they ventilated their homes.

Termites' lifestyles are very different from ants, bees, and wasps. The male survives with the queen. The young are workers. Some grow larger to become soldiers, whose job is to protect the nest, and a few may go on to start new colonies.

▶ The living space in the biggest termite nests is mainly underground. The tall spires provide ventilation which helps the termites control the temperature and humidity of their home.

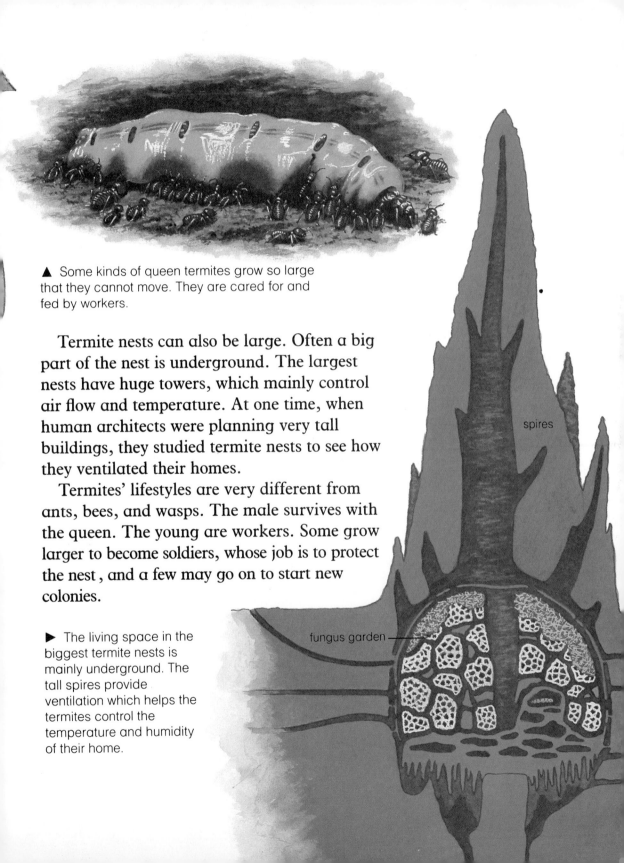

spires

fungus garden

Changing the Environment

We hear a lot about people harming the environment, and often forget that all living things alter their environment to some extent. But usually the changes are small and difficult to see. The most common cause of damage occurs when animals become too numerous. If they eat all the plants, the result may be soil erosion, and large areas of land can become useless. But not all alterations are harmful.

Rabbits can cause changes where they burrow, for deep-rooted plants will dry out and die. As the rabbits feed, they nibble all kinds of plants, so tree seedlings do not have the chance to grow.

▼ Rabbits can greatly damage the environment in which they live. The land around a rabbit warren is usually clear of all plants, except bitter or poisonous ones that the rabbits cannot eat.

Beavers also change their environment. They make lakes by building dams in fast-flowing streams. Generations of beavers may live by these lakes. They cut down all the nearby trees for food or for repairing the dam. Eventually the trees are too far away to be of any use, and the beavers move to a new site. But meanwhile the lake becomes filled with silt. Eventually the area flooded by the dam becomes a swamp, and then a dry meadow with a stream running through it. Settlers used to look for beaver meadows, because they were flat and fertile, a good area for growing crops.

▲ Beavers cut down trees to build dams and their homes.

▼ Most deer live and feed in forests. In some places where there are plenty of deer, the trees have been destroyed.

Animals In Our Homes

We think of our houses as belonging to us. But often they make good living places for other creatures as well. The outer walls are like cliffs, so some rock-dwelling creatures live there. Jumping spiders, normally at home on steep, rocky slopes, hunt resting insects on the walls of buildings. House martins and barn swallows make their mud nests on the small protected walls of barns and houses.

▼ The death watch beetle lives in old timbers in buildings. Other more harmful beetles sometimes survive in wood that has not been dried properly before being used.

▶ House martins build their mud nests on the walls of houses.

▶ The jumping spider catches insects on the outside walls of houses. Inside, other kinds of spiders catch insects that, for a short time, share our homes with us.

The wood used to build a house is sometimes attacked by scavenging insects, usually the grubs of wood-boring beetles. In the wild, they are recyclers, doing a valuable job of returning dead wood to the soil. They are unwelcome in our houses however, for we want the houses to stand, not to crumble. Most wood houses are now protected by powerful chemicals to deter these visitors.

Other insects may enter our homes. Moths and some beetles are often attracted by lights at night. They are harmless and can usually be put back into the open with no trouble. Those that can't escape may be caught by a spider lurking in a corner. Most spiders are useful killers of pests. Harmful species live in warm parts of the world, but few of these come into houses.

Parts of a house not often used, such as attics and cellars, may be taken over by small mammals. Most do no harm, but mice and rats may invade where they can find food.

Glossary

den a safe place within an animal's territory where it can sleep and rear its family.

dominant the leader within a group of animals. Within a group, the males and females will often have separate dominants.

environment the area within which an animal lives and all that it contains.

grub the larva or young of an insect. It is the stage of an insect's life after it hatches from the egg and before it develops into an adult. Caterpillars, beetles, flies, and bees are grubs when they hatch from their eggs.

habitat a living place such as a forest, a moor or a mountainside.

hibernation a sleep-like state during the cold part of winter in which the workings of the body slow down. When an animal hibernates it does not need to feed because it doesn't use much energy.

home range the area in which an animal lives and finds its food. It may overlap with home ranges of other animals of the same kind.

megapodes family name for a variety of birds found in Australasia and many South Pacific Islands.

nursery the place in which young animals are reared.

predator a hunter.

pupa the resting stage in an insect's life during which it develops from a grub into an adult. When there are more than one pupa, they are called pupae.

rear to breed and care for the young of a group.

territory part of an animal's living space that is defended against other members of the same species. It is usually used for protecting a food supply or for rearing young.

warren a group of rabbit burrows which are joined together by tunnels.

Index

Adelie penguins, *16*
antelopes, *18, 20*
ants, *22, 24, 25*
badgers, *11*
Baltimore orioles, *17*
Barn swallows, *28*
bats, *14, 18*
bears, *5, 12–13*
   brown, *12*
   polar, *12*
beavers, *27*
bees, *22–23, 25*
   bumblebee, *23*
   honeybee, *23*
   worker, *22*
beetles, *28, 29*
   death watch, *28*
   wood-boring, *29*
birds
   nests, *7, 14, 16–17*
   sea, *7, 16*
   song, *17*
black-footed ferret, *20*
boundaries, *5*
bullfinches, *17*
burrows, *13, 20–21*
butterflies, *4, 5*
caterpillars, *22, 23*
chimpanzees, *15*
coral reef, *8*
coteries, *20*
deer, *27*
dens, *10–11, 30*
   winter, *12–13*
displays, *5*
eagles, *16*
earwigs, *22*
elephants, *18*
environment, *4, 8, 26–27, 30*

finches, *17*
fish, *8–9, 16*
fortress, *10*
gannets, *7*
gorillas, *15*
grubs, *10, 22, 23, 24, 29, 30*
habitat, *8, 30*
hamsters, *13*
hedgehogs, *13*
hibernation, *13, 30*
home range, *4, 30*
houses, *28–29*
hummingbirds, *7, 16*
   hillstar, *6*
insects, *14, 16, 22–23, 24, 28, 29*
landmarks, *5*
mallee fowl, *17*
mammals, *14, 18–19, 20–21, 29*
marmots, *13*
martins, *19*
   house, *28*
megapodes, *17*
mice, *29*
   harvest, *18, 19*
mole rats, *21*
moles, *10*
moths, *29*
mound building birds, *17*
naked mole rats, *21*
nests, *7, 14–15, 16–17, 28*
nurseries, *18, 19, 23, 30*
orangutans, *15*
ovenbirds, *16*
owls, *6*
penguins, *16*
pine marten, *4, 18*
prairie dogs, *20, 21*

predators, *19, 21, 30*
pupae, *24, 30*
rabbits, *26*
   Old World, *19*
raccoons, *14*
rattlesnakes, *20*
reptiles, *16*
setts, *11*
shrews, *10*
songbirds, *17*
spiders, *16, 28, 29*
   jumping, *28*
squirrels, *14, 15, 20*
   gray, *14*
sticklebacks, *9*
swiftlets, *16, 17*
swifts, *17*
termites, *22, 24–25*
   queen, *25*
   soldiers, *25*
   workers, *25*
territories, *4, 5, 6–7, 10, 30*
   water, *8–9*
towns, *20–21*
tree homes, *14–15*
wading birds, *16*
wards, *20*
warrens, *19, 26, 30*
wasps, *22–23, 24, 25*
   caterpillar hunter, *23*
   social, *23*
   worker, *23*
weaverbirds, *16*
wild boar, *18*
woodpeckers, *14*
worms, *10*
yellow-shafted flicker, *5*
zebras, *20*